Environmental Refugees:
A Yardstick of Habitability

Jodi L. Jacobson

Worldwatch Paper 86
November 1988

Financial support for this paper was provided by the United Nations Fund for Population Activities. Sections of this paper may be reproduced in magazines and newspapers with acknowledgment to the Worldwatch Institute.

Printed on Recycled Paper

Table of Contents

Introduction

More than two years after an explosion at a nuclear reactor in the Ukraine spewed clouds of radiation from Kiev to Krakow, Soviet officials have announced plans to demolish the adjacent town of Chernobyl. This death warrant extinguished any hope of returning home for the city's 10,000 former residents. Because the world's worst nuclear disaster has permanently contaminated their home town, they will be forced to settle elsewhere.[1]

These people are refugees, though not by any standard definition. According to widely accepted doctrine, refugees are people who decide to seek asylum out of fear of political, racial, or religious persecution, or who leave their homes because of war or civil strife. This conventional notion, however, leaves out a new and growing class—environmental refugees.

Throughout the world, vast areas are becoming unfit for human habitation. These lands are being despoiled either through long-term environmental degradation or by brief but catastrophic events. Unsustainable land-use practices, for example, have reduced the ability of ecosystems to support life—called "carrying capacity"—throughout the Third World. On the other hand, high-risk technologies have sometimes resulted in accidents, such as the Chernobyl explosion, that leave whole regions uninhabitable for extended periods.

I would like to thank Hilary French for assistance with research, Susan Norris for production assistance, and the Worldwatch staff for their insightful comments on this paper.

6

The growing number of people fleeing from environmental decline adds a new dimension to an already controversial global refugee problem. The number of refugees in need of protection and assistance under traditional classifications, now more than 13 million, is mounting daily due to wars and insurrections, despotic governments, and deteriorating economic conditions, particularly in the Third World. Meanwhile, those nations that have been the traditional haven for refugees are increasingly trying to restrict this form of immigration.[2]

Most governments do not recognize environmental decline as a legitimate cause of refugee movements, choosing instead to ignore the issue. Neither the U.S. State Department nor the U.N. High Commissioner for Refugees, for example, collects data on this problem. Yet the number of environmental refugees—estimated by the author to be at least 10 million—rivals that of officially recognized refugees and is sure to overtake this latter group in the decades to come. The absence of public awareness and understanding of the decline in the earth's habitability only exacerbates the problem.

The degradation of agricultural land is the most widespread threat to habitability worldwide. But other factors, including the gradual poisoning of land and water by toxic wastes, and the effects of natural disasters made worse by human activity, are adding to the ranks of environmental refugees. Moreover, the expected rise of the sea level because of global warming threatens to reduce the planet's habitable area on a grand scale, perhaps forcing the evacuation of low-lying cities and agricultural land throughout the world.

To judge by what populations will put up with before they flee an environmental hazard, society's standards concerning habitability are fairly lax. People are willing to tolerate a broad range of threats to health and longevity. Witness the fact that, throughout the world, densely populated cities plagued by air and water pollution are the rule rather than the exception. And, in many countries, millions have built homes in areas prone to avalanches and floods.

"The number of environmental refugees rivals that of officially recognized refugees and is sure to overtake this latter group in the decades to come."

For every environmental refugee, then, there are thousands whose lives are compromised every day because of unhealthy or hazardous conditions. Because migration is a last resort—when conditions become so poor that life itself is in imminent danger—the rising number of environmental refugees should be seen as an important indicator of the extent and severity of worldwide environmental deterioration.

Long-term declines in habitability occur in stages, themselves often the result of imperceptible changes. After the initial stages, health may become threatened. Malnutrition is found throughout Africa and Asia where land degradation has cut crop yields. Increased cancer rates are associated with the high levels of toxic chemicals in towns along Louisiana's petroleum corridor.

In the final stages of decline, a town, city, or region may become virtually uninhabitable. For example, desertification—the impoverishment of land from human activities and natural stresses—has irreparably damaged millions of hectares of once-productive land and made refugees out of millions of sub-Saharan African farmers in this decade alone. Migration is the signal that land degradation has reached its sorry end.

Now it looks as if rising seas will supplant encroaching deserts and other forms of land degradation as the major threat to habitability in the not-too-distant future. Global warming, primarily the result of fossil fuel use in industrial countries, will hit developing nations the hardest. The Third World will suffer the greatest and most immediate impact of the rise in sea levels that will occur because of the thermal expansion of the oceans and the melting of the icecaps. The 1-meter increase projected over the next century will displace millions of people in the delta regions of the Nile and Ganges rivers, for instance, exacerbating land scarcity in the already densely populated nations of Egypt and Bangladesh. Protecting shorelines and wetlands, not to mention the infrastructure and water supplies of coastal cities, will require billions of dollars, perhaps even more than many well-off nations will be able to pay.

Unfortunately, while more and more land is rendered uninhabitable each year, population is increasing by 90 million annually—most of it in areas already in advanced stages of environmental degradation. All will need food, water, clothing, and shelter. One certain result of their struggle for these necessities will be further pressures on the earth's ecosystems and their ability to support human life.

In Search of Fertile Soils

In his landmark historical novel *The Grapes of Wrath*, John Steinbeck chronicled the economic and environmental disintegration of American farms in the Great Plains during the thirties. During that period, small farmers caught in the vise of poverty and debt took to planting crops from fencerow to fencerow to survive. Failed harvests and subsequent foreclosures rent the fabric of rural society, sending thousands of families westward in search of a livelihood. Most of these former landowners and sharecroppers became migrant workers.[3]

Though Steinbeck never used the term, his "Okies" were environmental refugees. To be sure, the Depression had a strong impact on U.S. agriculture. But it was the severely degraded environment dubbed the "Dust Bowl" that ultimately forced farmers from their land. Unsustainable farming practices had impoverished soils and made them more vulnerable to erosion by wind and rain. As the drought that became a hallmark of that decade deepened, rural peoples' economic margin of safety vanished along with their topsoil.

Today, this story is repeating itself in many parts of the world. Agricultural lands are degrading on every continent. This deterioration is most acute and its impact is greatest in those Third World countries where the majority of the people are farmers. Soil erosion

may cost Canada some $1 billion annually in reduced yields, but
Canadians do not starve. By contrast, sharply deteriorating land
resources in Africa imperil the lives of millions.[4]

Land degradation is most often associated with poverty. Indeed, the
two form a vicious circle. Throughout the Third World, subsistence
farmers eke out a living on land that is depleted of nutrients, stripped
of topsoil, and no longer able to withstand natural stresses such as
drought or heavy rain.

These cultivators are, in a sense, victims of circumstance. Though the
gross domestic product of most developing countries is dominated by
farm goods, few invest in the agricultural sector beyond their support
of cash crops for export. The low, government-regulated prices,
inadequate credit and extension services, and inequitable land tenure
characteristic of many developing countries have kept small land-
holders from increasing productivity in a sustainable manner. Eventually their land is depleted beyond restoration. Once this stage is
reached, people are forced to move.

Pressed by growing families and deepening poverty, farmers make
decisions to increase productivity that, in the long run, prove environ-
mentally and economically disastrous. Cultivating land that should
be fallowed, dividing already small plots among family members,
bearing numerous children to help with farm chores, cutting ever-
scarcer trees for fuel and fodder are all practices that, while they may
ensure a meager harvest for tomorrow, make certain famine is inevitable.

Mass migrations have become the enduring symbol of hunger. In
Ethiopia, relief workers watch the movements of villagers toward food
distribution centers as one indicator of conditions in the rural areas.
In the country's northern region, "stone deserts" have replaced nearly
4 million hectares of what was once fertile farmland. In June 1988, the
U.S. embassy in Addis Ababa reported that about 1 million people in

the highlands were about to move without warning due to famine conditions. Soil erosion and rapid loss of productivity ensure that the next drought will create a new wave of environmental refugees.[5]

Lacking official recognition, most environmental refugees go un-counted. Many fleeing land degradation are not classified as such because they simply move on to cultivate ever-more marginal lands. In Africa, thousands end up in the relief camps that are now regular fixtures on that continent. Others move to urban areas: the massive shift from rural regions to cities that has occurred in the Third World since mid century is due in large part to the complex of factors underlying land degradation.

"Throughout the Third World, land degradation has been the main factor in the migration of subsistence farmers into the slums and shantytowns of major cities, producing desperate populations vulner-able to disease and natural disasters and prone to participate in crime and civil strife," according to the United Nations Environment Pro-gram (UNEP). "Such exodus . . . exacerbate[s] the already dire urban problems. . . . And, at the same time, it has delayed efforts to rehabilitate and develop rural areas—through the lack of manpower and the increased negligence of land."[6]

Desertification, the most severe form of land degradation, is most acute in the arid and semi-arid regions. A UNEP survey has estimated that a total of 4.5 billion hectares around the world—fully 35 percent of the earth's land surface—are in various stages of desertification. These areas are home to more than 850 million people, many of whom are at risk of having their homes and livelihoods foreclosed by land degradation.[7]

Worldwide, desertification irretrievably claims approximately 6 million hectares each year. An additional 20 million hectares of productive rangelands and irrigated and rain-fed cropland are rendered perma-nently useless unless investments are made to restore them.[8]

> "Of all continents, Africa, a land where poor soils and variable rainfall pose a harsh climate for agriculture, has spawned the most environmental refugees."

About 135 million people inhabit areas undergoing such severe desertification. Soil scientist Harold Dregne of Texas Tech University notes that "50 million . . . have already experienced a major loss in their ability to support themselves. [Furthermore] an unknown percentage of that 50 million will have to abandon their agricultural way of life and join the overcrowded cities to seek relief."[9]

Of all continents, Africa, a land where poor soils and variable rainfall pose a harsh climate for agriculture, has spawned the most environmental refugees. Most come from the Sahel, a belt that spans several agroecological zones and stretches west to east across some nine countries from Mauritania and Senegal on into the Sudan. Desertification is accelerating in the Sahel, the world's largest area to be threatened by the wholesale loss of arable land. As the region's habitability declines, the movement of people increases: in the last 20 years the area's urban population has quadrupled.[10]

Two major droughts have occurred in the Sahel over the past two decades, one from 1968 to 1973 and the other from 1982 to 1984. The first signaled the beginning of Africa's dependence on outside assistance to feed its people. Even so, between 100,000 and 250,000 waited too long to migrate and died.[11]

To escape this fate, many Sahelians moved south and west to the coastal African nations. Whole villages were abandoned as the movement across and within borders got under way. The flux of environmental refugees was the largest ever witnessed: more than 250,000 people in Mauritania, 20 percent of its population, joined the already growing ranks of destitute farmers in the country's towns. Nearly 1 million environmental refugees in Burkina Faso (then Upper Volta), a sixth of the country's population, migrated to cities.[12]

By 1974 there were 200,000 people in Niger completely dependent on food distribution in towns and camps. In Mali, 250,000, or 5 percent of the population, were totally aid-dependent. The Ivory Coast, with a relatively stable and developed economy, became the principal destination for many refugees from the growing desert.[13]

Years of lower-than-average rainfall persisted throughout the seventies and eighties, leading up to the second drought. By early 1984, more than 150 million people in 24 western, eastern, and southern African countries were on the brink of starvation. By March 1985, the drought had forced an estimated 10 million people to abandon their homes in search of food.[14]

In just five countries—Burkina Faso, Chad, Mali, Mauritania, and Niger—more than 2 million people were displaced during this second drought. (See Table 1.) Some 3 million people, half the country's population, were affected by drought in Niger during 1983. There, Fulani and Tuareg pastoralists became paupers as two-thirds of their herds died. By June 1985, some 400,000 nomads had moved to the cities. In Mauritania the population of its capital, Nouakchott, swelled to four times its original size as desert swept the countryside. Since then, migration to Africa's cities has continued. United Nations estimates suggest that by the year 2000, Burkina Faso, Chad, Mali, Mauritania, Niger, and Senegal will have a collective urban population of 11.8 million, a 224-percent increase over 1975.[15]

The UNEP notes that desertification has worsened in the Sahel over the last decade, particularly in Chad, Mali, Mauritania, Niger, and Senegal, countries where low per capita incomes and rapid population growth compound rural poverty while large external debts and severe shortages of foreign exchange hamstring internal development efforts. Mauritania, one of the world's poorest countries, is the most threatened by desertification. Overgrazing by animals in its central and southern areas has stripped bare the vegetation that acted as a barrier to shifting sand dunes. Desertification has become so pervasive that dunes have covered whole villages and agricultural fields. Sand-choked schools, mosques, wells, and oases have been abandoned throughout the country. The ancient cities of Chinguetti, Tichitt,

Table 1: Displaced Population in Selected African Countries, September 1985

Country	People Displaced	Share of Population
	(number)	(percent)
Burkina Faso	222,000	3
Chad	500,000	11
Mali	200,000	3
Mauritania	190,000	12
Niger	1,000,000	16

Source: U.N. Office of Emergency Operations in Africa, *Status Report on the Emergency Situation in Africa As of 1 September 1985* (New York: United Nations, 1985).

Oualata, and Ouadane are under constant siege from glacier-like waves of sand. Mauritania's only major roadway, optimistically christened the "Highway of Hope," has become hopelessly impassable.[16]

Changes in land use under way since colonial times that have undermined the partnership between people and land are among the causes of the Sahel's decline. Both human and livestock populations have increased dramatically. One result has been a growing competition between farmers and livestock herders for scarce land. In the state of Borno in Nigeria, a stable agricultural balance had held for centuries due to complementary land-use patterns developed by the Hausa planters and the itinerant cattle-herders known as the Fulani. But a breakdown in land stewardship, combined with the region's droughts, has reduced productivity and heightened tensions between the groups. In many cases, the herders have had to migrate.[17]

Pastoralists, politically less powerful than their agrarian counterparts, often lose out in the struggle for land and are faced with the choice of grazing their herds on smaller patches or becoming settled farmers themselves. Long-accustomed to wandering the continent's arid and semi-arid lands in an ecologically balanced if somewhat tenuous partnership with nature, many of the great pastoral tribes, like the Fulani and the Tuareg, have been forced by government policy and land degradation to adopt a more sedentary lifestyle.

Governments throughout the Sahel encourage the establishment of cash-crop plantations and settled agriculture in rangelands for a host of economic and political reasons. But these types of farming, less ecologically appropriate in arid lands than pastoralism, reduce fallow periods and intensify degradation on land that is far too fragile for sustained cultivation. Not only have these policies further diminished the nomads' domain, but their increased conversion to sedentary farming concentrates large numbers of people and livestock around oases that in the past were visited only on a seasonal basis. This has had a devastating effect on local vegetation and water supplies. And because of the lower resiliency of arid ecosystems to environmental stress, seasonal pastureland around many oases has not recovered. Refugees from land degradation today, former nomads like the Tuaregs are sure to be on the move again within the next decade or so due to the declining habitability of the ecosystems they are crowding into.

Land degradation is also undermining habitability north of the Sahel. Larger human and cattle populations have exceeded the carrying capacity of arid lands in Algeria, Egypt, Libya, Morocco, and Tunisia. In Algeria, for example, desertification has begun to undercut the economy. The UNEP notes that north of the Saharan Atlas Mountains "some large settlements and cultivated lands are threatened by shifting sand dunes and sand drift." In Morocco, high population densities on arid lands are leading to desertification. Irrigation canals, roads, and oases are threatened by drifting sand and shifting dunes.[18]

In the southern part of the continent, deforestation, soil erosion, and the depletion of water supplies have driven tens of thousands of environmental refugees from their farmlands to other rural areas, into towns and cities, or into relief camps. In Botswana, for example, boreholes have been drilled in many parts of the country to supply the water needs of cattle herds that have multiplied in part due to government incentives to export beef. Water tables are now dropping steadily as a result, forcing herders to migrate or give up their stock.[19]

15

What seems to be self-reinforcing drought conditions have also taken hold in India. Between 1978 and 1983, western Rajasthan and parts of eastern India were gripped by serious drought. Thousands of farmers whose crops had failed for years on end began moving out of these areas by mid 1983 to neighboring Haryana and Madhya Pradesh. Many moved to the huge coastal city of Madras, where the influx caused lines for such basic commodities as water.[20]

In Latin America and the Caribbean, land degradation results from the combination of highly inequitable land distribution and rapidly growing populations. Latin America is home to some of the world's biggest cities in large part because of migration from rural areas. Millions of poverty-stricken farmers facing decades of agricultural neglect and land degradation throughout the mountains and the plains of South America fill the urban shantytowns of Sao Paulo, Rio De Janeiro, Mexico City, Lima, and La Paz. In many countries, particularly in Central America, the response to enduring poverty and environmental decline has been civil war and (often illegal) migration to the United States.[21]

Haiti, already the poorest country in the Western Hemisphere, also has the fastest-growing population. One-third of the nation's land, exhausted by decades of deforestation and poor husbandry, is now virtually useless, and about 40 percent of the population is malnourished. More than half of the land is held by less than 4 percent of the planters. The average holding for a peasant farmer is less than an acre.

When food production per person began to decline in the fifties, farmers began adding to their incomes by selling charcoal. As trees vanished, so did the topsoil, further reducing food production and increasing reliance on charcoal for income. A combination of political repression, economic decline, and environmental devastation has pushed an estimated 1 million refugees—one-sixth of Haiti's population—out of the country over the past decade.[22]

Agriculture is the backbone of developing economies. Yet throughout the Third World, farmers have been forced by financial and population pressures to adopt short-cut methods that are leading to long-term land degradation. By interfering with important natural cycles and overusing fragile, barely stable ecosystems, they are creating a self-reinforcing cycle of land deterioration. When the countryside is no longer able to produce a crop, the farmers along with the rest of the rural populace are forced to move on. Whether they end up in cities, relief camps, or cultivating marginal lands, these people constitute a growing class of environmental refugees.

Unnatural Disasters

When an earthquake in Colombia or a flood in India causes hundreds of deaths and leaves thousands homeless, society accepts these losses as unfortunate accidents of fate. These natural disasters are second only to land degradation as a factor in the growing number of environmental refugees. But more people are being killed or displaced by avalanches, cyclones, earthquakes, and floods than ever before, and close examination of the environmental backdrop against which these events occur suggests a strong human component.

"Unnatural disasters"—normal events whose effects are exacerbated by human activities—are on the rise. Human pressures on forests, soils, and land have rendered ecosystems less resilient, less able to cope with natural fluctuations. Ultimately, they collapse under other-

"Unnatural disasters are on the rise.
Human-induced changes in the
environment can turn a normal event
into a catastrophe."

wise normal stresses, creating and magnifying disasters such as landslides and floods.

At the same time, competition for land and natural resources is driving **17** more people to live in these marginal, disaster-prone areas, leaving them more vulnerable to natural forces. Hence, millions of Bangladeshis live on "chars," bars of silt and sand in the middle of the Bengal delta some of which are washed away each year by ocean tides and monsoon floods. Millions of Nepalis live in the areas most likely to be hit by earthquakes. And thousands of slum dwellers in the cities of Latin America perch on deforested hillsides prone to mudslides in heavy rain.

Human-induced changes in the environment can turn a normal event into a catastrophe. The deterioration of major watersheds in many Third World countries, for example, increases the number of devastating floods. The floods in Bangladesh and Sudan in 1988 are cases in point.

The flood in Bangladesh came on the heels of heavy monsoon rains in the Himalayan watershed of the Ganges River system. In addition to the 25 million left homeless, at least 1,200 people died and hundreds of thousands more contracted diseases as a result of contaminated food and water supplies. Weeks earlier and a continent away, the city of Khartoum was ravaged by the deepest flooding to strike the Sudan in this century. Some 1.5 million people were left homeless. In both cases, human-induced changes in the environment—deforestation, soil erosion, and disruption of the hydrologic cycle at the sources of the Ganges and the Blue Nile—helped to turn a heavy rain into a catastrophic flood.[23]

Bangladesh, with its per capita income of $160 among the world's poorest nations, is also one of its most densely populated. More than 110 million people—almost half the population of the United States— are packed into a country about the size of Wisconsin. Bangladesh sits on a vast, low-lying island of silt that makes up the world's largest river delta.[24]

Bangladeshis are accustomed to having water overflow the banks of their mighty rivers. Each year, monsoon rains quench the thirsty Indian subcontinent, shedding moisture essential for crops. Water accumulated in the Himalayan ranges of Bhutan, India, Nepal, and Tibet runs through the Ganges and two other large rivers, the Brahmaputra and the Meghna, into Bangladesh, supplying all but 10 percent of the country's water.[25]

This water is critical to Bangladesh's agricultural output: rice farmers, for example, depend on moderate annual floods for a successful harvest. And the silt carried by rivers and streams into the delta region helps maintain soil fertility. In moderation, these natural commodities are essential to Bangladesh; delivered in excess, they can be disastrous.

But up in the Himalayan watershed, where agrarian people depend on wood for fuel, a large and rapidly expanding rural population has been outstripping the carrying capacity of its environment. Deforestation, overgrazing, and unsustainable farming practices have diminished the soil's ability to absorb water. Available data, though not comprehensive, suggest that from one-half to three-fourths of the middle mountain ranges in Nepal and India have been deforested in the last four decades.[26]

In the past, truly massive floods hit Bangladesh only once every fifty years or so. But since mid century the number of large-scale floods has markedly increased. The country was heavily inundated on average once every four years through the seventies. After a flood in 1974, an estimated 300,000 people died in a famine that led to the overthrow of the country's founder, Sheik Mujibur Rahman. Since 1980, there have been several "fifty-year" floods, each worse than the last.[27]

Similarly, degradation of the Nile watershed contributed to the flooding in Sudan. The headwaters of the Blue Nile are in the highlands in Ethiopia, where a rich and diverse agriculture developed thousands of years ago. Today, the highlands constitute 90 percent of the arable **19** land, supporting 88 percent of the country's population and 60 percent of its livestock. But deforestation and poor soil husbandry, coupled with rapid population growth, have undermined the nation's agricultural base.[28]

Because land degradation has disrupted the hydrologic cycle, in which water is recycled to and from the atmosphere through soil absorption and plant transpiration, the entire region is considerably drier than in the past. Extensive floods are not normally a feature of the region downstream of the highlands, but in 1988 an exceptionally heavy rainfall, together with the watershed's reduced holding capacity, allowed a torrent to wreak havoc on the Sudanese plain below.

Some "natural" tragedies are the result of development strategies that blatantly disregard their impact on the environment. In 1983, a cyclone in the Philippines that normally might have caused fewer than one hundred fatalities killed thousands. Floods caused by the tropical storm were far more numerous and severe than in the past. "Villages built in places where flooding had not been a problem before are having to deal with patterns of water runoff that have been radically changed by the lumbering and mining operations which have spread unchecked," writes Debora MacKenzie in the *New Scientist*. "Slag heaps from mines have been thrown up in valleys, sometimes completely rerouting rivers, and whole forests have disappeared."[29]

Deforestation of mountains and hills that ring cities in the developing world has led to an increasing number of mudslides in urban shantytowns. In September 1987, more than 500 people were killed in a cascade of mud and rocks in Medellin, Colombia, after torrential rains had soaked the Andes for a week. Only half the population of Villa Tina, an impoverished suburb of Medellin, survived the catastrophe.[30]

Five months later a similar disaster hit the shantytowns of Rio de
Janeiro. Eighteen inches of rain fell on the city over a three-week
period, destabilizing the mountainsides once forested by soil-grabbing
trees but now blanketed with huts of scrap wood, adobe, and sheet
metal. Nearly 300 people died, 1,000 were injured, and more than
18,000 were left homeless. Mac Margolis of the *Washington Post* noted
that although this was the worst such storm since 1966, "Lately . . .
even modest rains have proven deadly . . . render[ing] this sophisti-
cated metropolis a hostage to the elements."[31]

These "unnatural disasters" are largely a product of the same kind of
land degradation discussed in the previous section, in which financial
and population pressures force both farmers and urban dwellers onto
marginal lands that soon lose their stability. But in this case the land
degradation—while devastating in itself and also to be feared because
it is self reinforcing—inhibits the ability of ecosystems to roll with
nature's punches. The result has been that the rare has become
commonplace, the extremes of weather that have been endured and
survived through the millennia are increasingly turning into full-
fledged catastrophes on a scale seldom before seen.

Home Is Where the Toxics Are

Although Guy Reynolds and his wife moved to Springfield, Vermont,
in a mobile home, they arrived with the intention of staying put. But
the Reynolds family and 59 other residents of the mobile-home park
in which they live have been forced to evacuate their residences. In
July 1988, the U.S. Environmental Protection Agency (EPA) ordered
their relocation when it determined that the park was situated atop a
landfill containing toxic chemicals. The Reynoldses and their neighbors are among a small but growing number of refugees from land
poisoned by hazardous wastes. Once confined to industrial countries,
the inherent conflict between disposal of toxic wastes and human
habitation is spreading around the world.[32]

Industrial society has developed a chemical dependency. More than 80,000 synthetic and organic compounds are commonly used worldwide in homes and by agriculture and industry. Modern chemicals usually offer an economic quick fix. Pesticides increase food production. Chemical-based plastics lower the cost of consumer items by replacing expensive metals. Almost every manufactured product and every manufacturing process involves the use of chemical compounds. Global production of organic chemicals alone increased from about 1 million tons a year in the thirties to about 250 million tons in 1985 and is now doubling every seven or eight years.[33]

But chemicals exact a high cost. Producing and using them often entails releasing large quantities of hazardous wastes into the environment. Residues from pesticides, herbicides, and fungicides end up in rivers, bays, and aquifers. Exhaust from cars and factories pollutes the air. Industrial wastes buried in landfills seep into groundwater and poison soils. Eventually, these waste products can affect biological productivity and human health. In some areas, the toxic hazards have forced people to move.

Chemical contamination can be sudden—the result of a rail accident, for instance. Or it can result from the insidious penetration of toxics into the atmosphere, the food chain, or water supplies. Although toxic wastes pose a pervasive threat to the environment, until recently few countries had laws regulating their disposal. As a result, many companies found it easier and cheaper to discard their wastes into landfills, waterways, or the atmosphere. The disposal of chemical wastes in landfills over the past several decades has created enormous problems for communities throughout the world that are faced with the choice of expensive cleanups or contamination of their environment by leaking toxics. Put another way, many of today's contaminated communities are being forced to pay for the toxic sins of the past.

In the United States, dumping wastes into landfills that were later topped off and used for other purposes, such as housing developments, became commonplace. Today, thousands of toxic waste sites are festering sores in towns and cities throughout the country, and a battle continues over the hazards they represent and who should bear the responsibility for cleaning them up. In some cases, people remain in their homes, accepting higher risks to health because they are unable to sell their property and cannot otherwise afford to move. In other cases, toxic contamination is so bad that whole communities become ghost towns virtually overnight.

Love Canal was one such community. Beginning in 1920, a partially completed channel between the upper and lower Niagara Rivers in upstate New York came into use as a municipal and chemical waste dump. Then, in 1953, the channel was filled in. Homes and schools were subsequently built on and around the site now known as Love Canal. Over time, chemicals buried in the canal began to surface, and residents often complained of the strange odors and substances emanating from the landfill.[34]

Bureaucratic and political waffling forestalled any investigation into the contents of the landfill until 1976, when a consultant discovered toxic chemical residues in the air and sump pumps of a high percentage of homes bordering the canal. High levels of carcinogenic polychlorinated biphenyls—PCBs—were found in the storm sewer system. Despite the investigation, little was done to protect the families of Love Canal.[35]

Not until 1978 did the New York State Department of Health, on the basis of evidence showing a high incidence of reproductive problems among women and high levels of chemical contamination in homes, the soil, and air, order the evacuation of pregnant women and of children under the age of two from 239 homes immediately surrounding the canal. Eventually, all but 86 of the 900 families living in Love Canal were evacuated. Purchasing the homes of the former residents cost the federal government $20 million. Another $200 million is being

"Today, thousands of toxic waste sites are festering sores, and in some cases contamination is so bad that whole communities become ghost towns."

spent to clean up the area. Recently, parts of the neighborhood were declared safe for habitation.[36]

Love Canal proved to be just the tip of the toxic iceberg. Thousands of other sites across the country are contaminated by both legally and illegally dumped wastes. Realizing that both government and industry would have to share the burden of the cleanup, the U.S. Congress enacted the Superfund program in 1980.[37]

Since then, a total of 1,390 families in 42 communities across the country have been relocated with Superfund money. In 1983, Times Beach, Missouri, a suburb of St. Louis with a population of 2,400, was abandoned and disincorporated as a result of the careless spraying on city streets of oil laced with highly toxic dioxin. A combination of federal and state funds helped pay for relocation of families in Globe, Arizona, and Centralia, Pennsylvania, in the same year.[38]

Because relocation is less costly and simpler than detoxifying contaminated sites—not to mention a faster way of protecting people—the U.S. government has increasingly used this option to deal with the toxic waste problem. Since 1985 this class of environmental refugees has more than doubled.[39]

Some who would flee remain in hazardous areas because of financial circumstances. About one-fifth of the petrochemical production in the United States is concentrated along the 85-mile stretch of the Mississippi River that winds from Baton Rouge to New Orleans, Louisiana. Local economies are primarily dependent on the jobs and income offered by the 135 chemical plants and seven oil refineries that line this corridor. But the region absorbs more toxic substances annually than do most entire states, including such dangerous substances as vinyl chloride, a carcinogen and suspected embryotoxin.[40]

According to the *Washington Post*, "The air, ground, and water along this corridor are so full of carcinogens, mutagens, and embryotoxins that an environmental health specialist defined living [there] as 'a massive human experiment,' the state attorney general called the pollution 'a modern form of barbarism,' and a chemical union leader now refers to it as 'the national sacrifice zone.' " Several towns in the corridor exhibit uncommonly high rates of cancer and miscarriages.[41]

Because most of these compounds represent creeping rather than sudden dangers to health, and because little research has been undertaken to separate out the contribution of toxic chemicals from other health threats, government protection or assistance for relocation does not extend to residents of Louisiana's chemical corridor. Indeed, public policies—such as lax controls on polluters—have encouraged the growth of this industry. As a result, hundreds of thousands of people remain subject to the dangers of toxic poisoning and disease in an area that is barely inhabitable.

Urban residents around the world have long tacitly accepted the reality of living with higher levels of pollution in their immediate environment, particularly in the air they breathe. Automobiles, power plants, and industrial plants are the biggest contributors to air pollution. Where pollution control technology is unavailable and regulations are unenforced, as in Eastern Europe and the Soviet Union, regions where the post-war rush to industrialize was given precedence over environmental protection, emissions have made atmospheric pollution so bad as to render whole regions virtually uninhabitable.

Pollution poses grave threats to agriculture and human health throughout the Eastern European nations. The Polish government, for example, recently declared the village of Bogomice and four others "unfit for human habitation" due to the extremely high levels of heavy metals in the air and soil deposited by emissions from nearby copper smelting plants. The government is encouraging villagers from this region to resettle elsewhere by offering compensation.[42]

> "Sudden accidents, such as a rail crash, fire, or explosion, can instantaneously confer upon thousands of people the status of environmental refugee."

Likewise, in the Soviet Union, the quality of air, soil, water, and forest resources is in rapid decline. Fyodor Morgun, head of the country's Environmental Protection Committee, declared recently that degradation from industrial waste has reached the proportions of "a Chernobyl-like catastrophe." Problems are particularly acute in the Ural Mountains, a region of heavy industry. In December 1987, *Pravda* stated that the industrial city of Ufa, with a population of nearly 1 million, had become "unfit for human habitation."[43]

Sudden accidents, such as a rail crash, fire, or explosion, can instantaneously confer upon thousands of people the status of environmental refugee. In the 19 years leading up to 1978, there were seven major chemical accidents worldwide, killing a total of 739, injuring 2,647, and forcing 18,230 from their homes. All but one occurred in industrialized countries.[44]

Among the worst examples of an accident in the industrial world was the 1976 explosion at a small chemical plant in Seveso, Italy, that sent a cloud of smoke and highly toxic dioxin particles wafting over the countryside. Estimates of the amount of dioxin released range up to 5 kilograms, though to this day no one is completely certain. Eight hundred people were evacuated from their homes for more than a year. Although many have returned, questions about their health linger.[45]

About 200 people, mainly children, were struck by chloracne, a severe rash, as a result of exposure to the chemical. Incomplete data suggest an elevated rate of birth defects in the two years following the explosion. Whether the accident will result in a higher incidence of cancers with long latency periods will only be known after time passes.[46]

Since 1978, the number and severity of toxic disasters has increased, with more of these in the Third World. In the next eight years, there were 13 major chemical accidents. The numbers tell the tragedy: 3,930 dead, 4,848 injured, and nearly 1 million evacuated.[47]

Higher wage costs and tighter controls on production and disposal of hazardous chemical materials in industrial nations, along with the development of a global market for chemical products, have sent some multinational firms scurrying to build plants in developing nations. Experience shows that such investments can be a mixed blessing. Although they gain some jobs and revenue from the chemical industry, most developing countries have neither laws controlling toxic chemicals nor the technical and institutional capacity to put them into force. The general lack of controls makes incidents of contamination more likely.[48]

The toxic leak in 1984 at Bhopal, India, was perhaps the worst example of this trend. A Union Carbide pesticide plant accidentally released a cloud of deadly methyl isocyante over the town, killing about 2,500 and sending more than 200,000 fleeing for their lives. As many as 100,000 people are still suffering side effects, such as blurred vision, disabling lung diseases, intestinal bleeding, and neurological and psychological disorders. Bhopal, less a case of permanently reduced environmental habitability, is certainly evidence of the dangers inherent in many of the industrial and development choices being made today.[49]

Nuclear reactor accidents have the most pervasive and long-lasting consequences of any industrial catastrophe. In April 1986, an explosion and fire at the Chernobyl nuclear power plant in the Soviet Ukraine caused a partial meltdown of the plant's reactor core. More than 7 tons of radioactive material was hurled into the atmosphere, eventually contaminating land, food, and water throughout much of Europe. Twenty-eight people died as a result of acute radiation poisoning within 75 days of the accident, while another 300 were treated for serious radiation exposure. More than 100,000 people were evacuated

from their homes and an area up to 2,600 square kilometers was
rendered uninhabitable.[50]

Chernobyl was the worst reactor accident in history and will not easily
be forgotten. But the fact remains that an even more serious disaster
could occur at any time at one of the reactors in a densely populated
area. Assessing the risks inherent in using nuclear power under the
most optimistic conditions, nuclear analysts have determined that with
500 nuclear plants in operation, there would be one core-damaging
accident every 20 years, based on one accident for every 10,000 years
of reactor operation. But the Three Mile Island accident in the United
States occurred after only 1,500 years of worldwide reactor operation,
and Chernobyl occurred after another 1,900 years.[51]

In Worldwatch Paper 75, *Reassessing Nuclear Power: The Fallout From
Chernobyl*, Christopher Flavin writes: "With the overall size of the
nuclear industry continuing to grow rapidly, the chances and likely
frequency of a serious accident are increasing as well. If the current
accident rate were to continue, there would be three additional
accidents by the year 2000. At that point, with 500 reactors in
operation, core-damaging accidents would occur every four years." A
Chernobyl-like accident at the Indian Point plant in the New York
metropolitan region could require the permanent evacuation of more
than 1 million people.[52]

The rapidly growing volume of hazardous waste in industrialized
nations, coupled with high disposal costs, has led some companies to
export their industrial residues to the Third World, threatening a new
wave of chemical illnesses and refugees. It costs from $250 to $350 per
ton to dispose of hazardous municipal and industrial wastes in the
United States, for example, but some developing countries will accept
such wastes for as little as $40 per ton.[53]

Local conditions and lack of monitoring or waste treatment means that
a large proportion of these imported wastes will end up in the local

environment. Frequent rains and poor soils in tropical areas hasten the migration of chemical wastes into groundwater supplies.

Thousands of tons of U.S. and European wastes have already been shipped to Africa and the Middle East. The 3,800 tons of toxic waste from Italy illegally dumped in the small Nigerian port town of Koko in five shipments between 1987 and 1988 contained at least 150 tons of PCBs—the chemical that put Love Canal on the map. The contamination was the result of an illegal deal between an Italian waste contractor and several corrupt Nigerian government officials. Residents of Koko complained that the odors given off by the leaking sacks and containers of waste made them ill. The government plans to evacuate the 5,000 residents from their homes. The Italian government has since accepted responsibility for reclaiming and destroying the wastes but, as of this writing, a ship laden with the toxics from Koko had yet to find a port that would accept the hazardous chemicals.[54]

Similarly, waste shipped from Italy to Lebanon in May 1988 will be reclaimed by Italy and incinerated aboard a ship in the middle of the Pacific. More than 2,400 tons of toxic chemicals found their way to Beirut in a transaction involving both Lebanese and European merchants, without government supervision. Lebanon's health minister recognized the threat immediately: these wastes are "poisonous and harmful to man and the environment," he said. Other countries, such as Benin, the Congo, and Guinea-Bissau, are reconsidering plans to accept large quantities of waste from the United States and Europe in the wake of an international outcry against waste export, thus averting the environmental and human catastrophes that would likely ensue. The Congo, for example, recently canceled a contract to accept 20,000 to 50,000 tons of pesticide residue and sludge waste a month from a firm in New Jersey.[55]

Growing recognition and control of hazardous waste disposal may make the world a safer place to live—for a while. Two international treaties that would curtail trade in hazardous wastes are being negotiated, and individual countries and regions are beginning to take action on their own. In May 1988, the Organization of African Unity passed a resolution condemning the practice of accepting toxic wastes from the industrial world, and various of its member states are taking action to implement the declaration. The same month, the European Parliament urged its governments to adopt national legislation that would ensure recipient countries can handle the wastes. In the United States, two bills have been introduced in the Congress that would restrict the export of these hazardous substances. Despite such efforts, little has been done to reduce at the source the large volumes of toxic substances produced by industrial activity.[56]

The Threat of Inundation

Around the world, the Dutch are perhaps best known for their achievements in water engineering. And well they should be: without the carefully maintained stretches of dikes (400 kilometers long) and sand dunes (200 kilometers) built by Holland's engineers to hold back the sea, more than half of the country would be uninhabitable. Currently, 8 million people make their homes and livelihoods on this reclaimed delta, a region of unquestioned social and economic importance. The stakes in protecting it, however, are certain to mount. Rising sea levels promise to test even the Netherlands' capabilities in water management.[57]

Among the various environmental problems that cause the displacement of people from their habitats, none rivals the potential effects of sea level rise as a result of human-induced changes in the earth's climate. A 1-meter rise in ocean levels worldwide, for example, may result in the creation of 50 million environmental refugees from various countries—more than triple the number in all recognized refugee categories today.

Most scientists are in agreement that a global warming is under way, caused by the accumulation of "greenhouse gases" due primarily to fossil fuel use in industrial countries. The uncertainties lie in just how much higher the earth's average temperature will go, and how quickly the increase will take place. Recent estimates predict that a global temperature increase of 1.5 to 4.5 degrees Celsius can be expected as early as 2030.[58]

Small increases in the earth's average temperature due to the greenhouse effect will lead to a rise in the global sea level for two reasons. First, as atmospheric temperatures rise, so too will the average temperature of the oceans. Because of this heat transfer, the waters of the earth will expand. Second, higher temperatures will also cause polar ice to melt, further raising the level of the sea. Large masses of ice, such as the West Antarctic sheet, may break off, displacing water and further raising the sea level.[59]

If correct, the predicted temperature changes would precipitate a rise in sea level of 1.4 to 2.2 meters by the end of the next century. (In comparison, global sea level rise has probably not exceeded 15 centimeters over the past century.) Such an increase will affect people and infrastructure around the globe. Yet, while sea level rise due to global warming is induced largely by the industrial world, developing countries stand to suffer the most immediate and dramatic impacts.[60]

President Maumoon Abdul Gayoom of the Maldives put it bluntly when he stated: "The predicted effects of the change are unnerving. There will be significant shoreline movement and loss of land. A higher mean sea level would inevitably . . . increase frequency of inundation and exacerbate flood damage. It would inundate fertile deltas, causing loss of productive agricultural land and vegetation, and increase saline encroachment into aquifers, rivers, and estuaries. The increased costs of reconstruction, rehabilitation, and strengthening of coastal defense systems could turn out to be crippling for most affected countries."[61]

> "Among the various environmental problems that cause the displacement of people, none rivals the potential of sea level rise as a result of climate change."

Some climatologists now estimate that the rate of sea level rise will accelerate after 2050, reaching 2 to 3 centimeters a year. But actual sea level rise will be much higher in some regions than others because of obvious differences like land elevation and because of less obvious differences in geological processes such as tectonic uplift or subsidence in coastal areas. Subsidence is a key issue in the case of deltas, such as the Nile and Ganges river deltas.[62]

Under natural conditions, deltas are in a state of dynamic equilibrium, forming and breaking down in a continuous pattern of accretion and subsidence. The Mississippi River delta in the United States, for example, was built up by sediments deposited during floods and laid down by the river along its natural course to the sea. Over time, these sediments accumulate. But regional and local tectonic effects, along with compaction, cause the land created to subside. Local subsidence alone can be as great as 10 centimeters per year. Local rates of sea level rise, then, depend on the sum of global sea level rise and local subsidence.[63]

Subsidence is likely to accelerate where subterranean stores of water or oil are tapped. In Bangkok, Thailand, local subsidence has reached 13 centimeters per year due to a drop in the water table caused by excessive withdrawals of groundwater over the past three decades. Moreover, channeling, diverting, or damming rivers can greatly reduce the amount of sediment that reaches a delta. When this happens, as it has on the Mississippi and the Nile, sediment accumulation will not offset subsidence, resulting in heavier shoreline erosion and an increase in water levels.[64]

Low-lying delta regions, important from both an ecological and social standpoint, will be among the first land areas lost to inundation under even slight rises in sea level. Fertile deltas are important sources of food and other products. Of the world's major deltas, several, including the Bengal and Nile, are also densely populated. As a result, these regions will be the single greatest source of refugees from sea level rise.

A recent study by researchers at Woods Hole Oceanographic Institute in Massachusetts showed the combined effects of sea level rise and subsidence on Bangladesh and Egypt, where the homes and livelihoods of some 46 million people are potentially threatened.[65]

The researchers, headed by senior scientist John Milliman, developed three possible scenarios under two estimates of sea level rise: a minimum of 13 centimeters by 2050 and 28 centimeters by 2100, and a maximum of 79 centimeters by 2050 and 217 centimeters by 2100. Under the "best case" scenario, the researchers assume the minimum rise in global sea level and a delta region in equilibrium. The second scenario, called the "worst case," assumes the maximum rate of sea level rise and the complete damming or diversion of the river system draining into the delta. In this case, the rate of natural subsidence must then be added to the absolute rise in sea level.[66]

The third scenario is referred to as the "really worst case." It assumes that excessive groundwater pumping from irrigation and other uses accelerates natural subsidence. To calculate the economic implications of these three cases on both Egypt and Bangladesh, Milliman and his colleagues assumed present day conditions, such as the estimated share of total population now living in areas that would be inundated and the share of economic activity that is derived from them. Continued settlement and population growth in these areas will of course make for even more future environmental refugees.[67]

Although most of Bangladesh's population already lives at the margin of survival and can ill afford to deal with additional environmental disruptions, the country stands to be severely affected by any rise in sea level.[68]

The Bengal delta, resting at the confluence of the Ganges, Brahmaputra, and Meghna rivers, is the world's largest such coastal plain and comprises about 80 percent of the country's total area (about one-fifth is water). As a result, the nation's inhabitants are subject to annual floods both from the rivers and from ocean storm surges.[69]

Just how severely sea level rise will affect Bangladesh depends in part on the pace at which damming and channeling of rivers proceeds and the rate of groundwater withdrawal. Although annual flooding is severe and can damage up to one-third of the crops grown on the flood plains, large areas of the delta region suffer rain deficits for the rest of the year, thus creating a large incentive to divert river water for agriculture. Such diversions would impact heavily on the amount of sediment available to offset natural subsidence.[70]

On the basis of the limited data available, the Woods Hole researchers have concluded that the increasing withdrawal of groundwater in Bangladesh is affecting subsidence rates. Between 1978 and 1985, there was at least a six-fold increase in the number of wells drilled in the country; more than 100,000 shallow tubewells and 20,000 deeper ones were counted. Sediment samples suggest that groundwater withdrawal may have raised subsidence to at least twice the natural rate.[71]

Taking these factors into account, Milliman and his colleagues estimate that the sea level will rise as much as 209 centimeters along the Bangladesh coast by 2050. Half the country lies at elevations of less than 5 meters. Loss of land under the 13-centimeter rise in the best case scenario is minimal—less than 1 percent of the country's total. In the 144-centimeter rise in the worst case, however, 16 percent of the nation's land would be lost. In the really worst case, local sea level rise would be 209 centimeters and 18 percent of the habitable land would be under water. (See Table 2.) As a result, more than 17 million people would become environmental refugees.[72]

By the year 2100, the really worst case scenario shows that 35 percent of the nation's population—some 38 million people—will be forced to relocate. The economic effects will be widespread: Because 31 percent of Bangladesh's gross national product is realized within the land area

Table 2: Effects of Sea Level Rise on the Bengal and Nile River Deltas Under Two Scenarios

	Sea Level Rise					
	Global Sea Level Rise	Local Land Subsidence	Local Sea Level Rise	Loss of Habitable Land	Population Displaced	Loss of Gross National Product
	(centimeters)				(percent)	
Bangladesh 2050						
Worst Case	79	65	144	16	13	10
Really Worst Case	79	130	209	18	15	13
Bangladesh 2100						
Worst Case	217	115	332	26	27	22
Really Worst Case	217	230	447	34	35	31
Egypt 2050						
Worst Case	79	22	101	15	14	14
Really Worst Case	79	65	144	19	16	16
Egypt 2100						
Worst Case	217	40	257	21	19	19
Really Worst Case	217	115	332	26	24	24

Source: John D. Milliman et al., *Enviromental and Economic Impact of Rising Sea Level and Subsiding Deltas: The Nile and Bengal Examples*, Woods Hole Oceanographic Institution, Woods Hole, Massachusetts, unpublished paper, 1988.

that will be lost, an already poor country will have to accommodate its people on a far smaller economic base.[73]

Bangladesh will suffer further from the likely destruction of coastal mangrove forests upon which 30 percent of the country's population depends to some extent. Extensive river diversions will markedly decrease the amount of freshwater discharged into this coastal envi-

"Moving farther inland, millions of
environmental refugees will have to
compete with the local populace for
scarce food, water, and land."

ronment, while higher sea levels will increase saltwater, thus reducing mangrove forest cover and disrupting major fisheries within this fragile ecosystem. Reductions in river outflow may also accelerate the intrusion of brackish water into aquifers, a problem that already extends 240 kilometers inland.[74]

The combined effects of warmer climates and higher seas will make tropical storms more frequent and more destructive worldwide, raising the toll in lives and further decreasing the habitability of coastal areas. Milliman notes that cyclones originating in the Bay of Bengal before and after the rainy season already devastate the southern part of Bangladesh on a regular basis. An average of 1.5 severe cyclones now hit the country each year. Storm surges as much as 6 meters higher than normal can reach as far as 200 kilometers inland. Total property loss from storms in the region between 1945 and 1975 has been estimated at $7 billion, and some 300,000 lives were lost in 1970 when surge waters covered an estimated 35 percent of Bangladesh's land area. The impact of stronger and more frequent storms on this densely packed country is unthinkable.[75]

Where will those displaced by rising seas go? Moving farther inland, millions of environmental refugees will have to compete with the local populace for scarce food, water, and land, perhaps spurring regional clashes. Moreover, existing tensions between Bangladesh and its large neighbor to the west, India, are likely to heighten as the certain influx of environmental refugees from the former rises. Eventually, the combination of rising seas, harsher storms, and degradation of the Bengal delta may wreak so much damage that Bangladesh as it is known today may virtually cease to exist.

Egypt's habitable area is even more densely populated than Bangladesh. By and large, Egypt is desert: less than 4 percent of the country's land is cultivated and settled, leading to a population density of 1,800 people per square kilometer in the settled region. The Nile river and its delta, accounting for nearly all of the country's productive land, is Egypt's economic lifeline.[76]

Damming has virtually reduced the Nile's contribution of sediment and fresh water to the Mediterranean to a trickle. Milliman's study points out that because the Nile has been dammed, only the "worst" and "really worst" cases are relevant for Egypt, since most of the sediment that would offset subsidence of the delta is trapped upstream. Consequently, sea level rise would range between 101 and 144 centimeters by 2050, rendering up to 19 percent of Egypt's already scarce habitable land unlivable.[77]

If the higher increment were realized, more than 8.5 million people would be forced to relinquish their homes to the sea and Egypt would lose 16 percent of its gross national product. By 2100, sea level rise will range between 257 and 332 centimeters, inundating up to 26 percent of habitable land and affecting an equal percentage of both population and domestic economic output. Several shallow, brackish lakes along the coast, accounting for 50 percent of the nation's fish catch, would also be endangered.[78]

While neither Bangladesh nor Egypt is likely to markedly influence global emissions of greenhouse gases or sea level rise, they each wield considerable control over local sea levels. Development policies chosen in the near future will determine the rates of degradation and subsidence of these respective deltas. Milliman puts it well: "Saving crops and lives on the short-term basis . . . may lead to a long-term loss of land (and at least indirectly lives)."[79]

In 2100, cartographers will likely be drawing maps with new coastlines for many countries as a result of sea level rise. They may also make an important deletion: by that year, if current projections are borne out, the Maldives may have been washed from the earth. The small nation, made up of a series of 1,190 islands in atolls, is nowhere higher in elevation than 2 meters. A mean sea level rise of equal height would

submerge the entire country. With a 1-meter rise, well within the
expected increment of the next century, a storm surge would, in the
words of President Gayoom, be "catastrophic and possibly fatal."
Other such endangered places include the Pacific islands of Kiribati,
Tuvalu, and the Marshalls.[80]

Developed nations, heavily reliant on the burning of fossil fuels over
the past century, must assume the primary responsibility for global
warming and its consequences. And while they are in a far better
financial position than developing countries to undertake the remedial
technological measures necessary to save coastal areas and inhabited
land (thereby mitigating the problem of environmental refugees) these
actions will cost them dearly. The Netherlands, for example, will have
to spend at least $5 billion by 2040 shoring up dikes and increasing
drainage capacity to save their delta region. Large though these
expenditures are, they are trivial compared with what the United
States, with more than 19,000 kilometers of coastline, will have to
spend to protect its territorial integrity.[81]

Conclusion

On every continent, the living patterns of people are at odds with
natural systems. In industrial countries, consumption patterns and
industrial policies that ignore environmental limits have fouled the
planet with everything from toxic wastes to greenhouse gases. In the
Third World, population growth, poverty, and ill-conceived development policies are the root cause of environmental degradation. The
large and growing number of refugees worldwide that has resulted
from these trends is living evidence of a continuing decline in the
earth's habitability.

Environmental refugees have become the single largest class of displaced persons in the world. They fall into three broad categories:
those displaced temporarily because of a local disruption such as an

avalanche or earthquake; those who migrate because environmental degradation has undermined their livelihood or poses unacceptable risks to health; and those who resettle because land degradation has resulted in desertification or because of other permanent and untenable changes in their habitat. Although precise numbers are hard to fix due to lack of data, it appears that this last group—the permanently displaced—is both the largest and the fastest growing.

Until sea level rise overtakes it, land degradation will remain the single most important cause of environmental refugees. Land degradation occurs in stages, moving from moderate to severe desertification, for example. Thus it produces refugees in both of the last two categories. Refugees from land degradation often migrate from region to region cultivating one plot of marginal land after another, exacerbating the problem and moving on when the land no longer produces enough to meet basic needs. Others fill Third World cities. Across Africa, Asia, and Latin America, the swelling ranks of such refugees are signaling a large scale decline in the carrying capacity of agricultural lands. At some point this land will become so damaged that it is not economically or technically feasible to restore it to original levels of productivity.

Current trends are likely to worsen over the next few decades unless society acts to combat the problems underscored by the creation of environmental refugees. More and more land will be rendered unproductive or uninhabitable, whether through desertification, toxic pollution, or unnatural disasters. By the middle of the next century, the combined number of environmental refugees from all these and the inevitable rise in sea levels because of global warming will probably exceed the number of refugees today from all other causes by a factor of six.

The vision of tens of millions of persons permanently displaced from their homes is a frightening prospect, one without precedent and likely to rival most past and current wars in its impact on humanity. The growing number of environmental refugees today is already a rough indicator of the severity of global environmental decline. This yardstick may be imprecise but its message could not be clearer.

Notes

1. "Plan to Raze Chernobyl Reported," *New York Times*, October 9, 1988.

2. U.S. Committee For Refugees, *World Refugee Survey, 1987 in Review* (Washington, D.C.: American Council for Nationalities Service, 1988).

3. John Steinbeck, *The Grapes of Wrath* (New York: Viking Press, 1939).

4. Standing Committee on Agriculture, Fisheries, and Forestry, *Soil at Risk: Canada's Eroding Future*, A Report on Soil Conservation to the Senate of Canada (Ottawa: 1984).

5. Paul Harrison, *The Greening of Africa* (New York: Penguin Books, 1987); Robert M. Press, "Ethiopia Appears to Stay One Small Step Ahead of Famine," *Christian Science Monitor*, June 15, 1988.

6. Essam El-Hinnawi, *Environmental Refugees* (Nairobi: United Nations Environment Program (UNEP), 1985).

7. UNEP, *General Assessment of Progress in the Implementation of the Plan of Action to Combat Desertification* (New York: United Nations, 1984); figure on earth's total land surface from James H. Brown, *Biogeography* (St. Louis: C.V. Mosby Company, 1983); UNEP, "Sands of Change: Why Land Becomes Desert and What Can be Done About It," *UNEP Environment Brief No. 2*, 1988; H.E. Dregne and C.J. Tucker, "Desert Encroachment," *Desertification Control Bulletin*, No. 16, 1988. See also Peter D. Little, ed., *Lands at Risk* (Boulder, Colorado: Westview Press, 1987); M.B.K. Darkoh, "Socio-economic and Institutional Factors Behind Desertification in Southern Africa," *Area*, Vol. 19, No. 1, 1987.

8. UNEP, "Sands of Change."

9. Ibid.; H. E. Dregne, *Desertification of Arid Lands* (New York: Harwood Academic Publishers, 1983).

10. UNEP, "Sands of Change"; for information on the ecology of the Sahelian region see Harrison, *Greening of Africa.*; James Brooke, "Some Gains in West Africa's War on the Desert," *New York Times*, September 13, 1987.

11. UNEP, "Sands of Change"; Brooke, "Some Gains in West Africa's War."

12. UNEP, "Sands of Change"; Carolyn M. Somerville, *Drought and Aid in the Sahel* (Boulder, Colorado: Westview Press, 1986); Anders Wijkman and Lloyd

Timberlake, *Natural Disasters: Acts of God or Acts of Man?* (Washington, D.C.: International Institute for Environment and Development, 1984.)

13. Wijkman and Timberlake, *Natural Disasters;* UNEP, *Environmental Refugees.*

14. UNEP, *Environmental Refugees;* Harrison, *Greening of Africa.*

15. Harrison, *Greening of Africa;* Wijkman and Timberlake, *Natural Disasters;* UNEP, "Sands of Change." See also William S. Ellis, "Africa's Sahel: The Stricken Land," *National Geographic,* August 1987.

16. UNEP, "Sands of Change"; Sidy Gaye, "Glaciers of the Desert," *Ambio,* Vol. 16, No. 6, 1987.

17. Oakland Ross, "Where the Water Was," *Globe and Mail,* April 23, 1988.

18. UNEP, *Desertification Control in Africa: Actions and Directory of Institutions* (Nairobi: UNEP, 1985).

19. Harrison, "Greening of Africa"; Anne Charnock, "An African Survivor," *New Scientist,* July 3, 1986; Susan Ringrose and Wilma Matheson, "Desertification in Botswana: Progress Toward a Viable Monitoring System," *Desertification Control Bulletin,* No. 13, 1986.

20. Mary Anne Weaver, "India: The 'Greening' of a Bad Drought," *Christian Science Monitor,* May 25, 1983; UNEP, "Sands of Change"; J. Bandyopadhyay and Vandana Shiva, "Drought, Development, and Desertification," *Economic and Political Weekly,* August 16, 1986; Jayanta Bandyopadhyay, "Political Ecology of Drought and Water Scarcity," *Economic and Political Weekly,* December 12, 1987; Steven R. Weisman, "India's Drought Is Worst in Decades," *New York Times,* August 16, 1987; Anthony Spaeth, "Harshest Drought in Decades Devastates India's Crops, Slows Economic Growth, *Wall Street Journal,* August 19, 1987.

21. Lester R. Brown and Jodi L. Jacobson, *The Future of Urbanization: Facing the Ecological and Economic Constraints,* Worldwatch paper 77 (Washington, D.C.: Worldwatch Institute, May 1987).

22. Mark Kurlansky, "Haiti's Environment Teeters On the Edge," *International Wildlife,* March-April 1988; U.S. Committee for Refugees, *World Refugee Survey.*

23. Richard M. Weintraub, "Flooding Worsens in Bangladesh," *Washington Post*, September 5, 1988; "Bangladesh Intensifies Appeal for Flood Aid," *New York Times*, September 4, 1988; Blaine Harden, "Nile Floods in Sudan Termed Record, Could Deepen," *Washington Post*, August 10, 1988; Narul Huda, "Bangladesh Blames Neighbors for its Floods," Energy and Environment Group Features, New Dehli, 1987.

24. Bangladesh's per capita income figure from World Bank, *World Development Report 1988* (New York: Oxford University Press, 1988); Population Reference Bureau (PRB), "1988 World Population Data Sheet," (Washington, D.C.: 1988); "Life in Bangladesh Delta: A Race Bred By Disaster," *New York Times*, June 21, 1987.

25. Huda, "Bangladesh Blames Neighbors."

26. International Task Force, *Tropical Forests: A Call to Action, Part 2: Case Studies* (Washington, D.C.: World Resources Institute, 1985).

27. Huda, "Bangladesh Blames Neighbors"; "Misery Rising in the Floods in Bangladesh," *New York Times*, September 5, 1987; S. Kamaluddin, "Flood of Woes," *Far Eastern Economic Review*, November 8, 1984.

28. Hans Hurni, "Degradation and Conservation of the Resources in the Ethiopian Highlands," *Mountain Research and Development*, Vol. 8, Nos. 2–3, 1988.

29. Debora MacKenzie, "Man-made Disaster in the Philippines," *New Scientist*, September 13, 1983.

30. "120 Die in Avalanche of Mud in Colombian Slum," *New York Times*, September 29, 1987.

31. Mac Margolis, "Rio's Mudslides Partly Self-inflicted," *Washington Post*, February 28, 1988.

32. Sally Johnson, "Toxic Waste Uprooting Elderly From Trailer Park," *New York Times*, July 17, 1988. For information on hazardous waste problems in the developing world, see H. Jeffrey Leonard, "Hazardous Wastes: The Crisis Spreads," *National Development*, April 1986.

33. UNEP, "Hazardous Chemicals," *UNEP Environment Brief No. 4*, Nairobi, 1988. See also Sandra Postel, *Defusing the Toxics Threat: Controlling Pesticides and Industrial Waste*, Worldwatch Paper 79 (Washington, D.C.: Worldwatch Institute, September 1987).

34. Love Canal Homeowners Association, *Love Canal: A Chronology of Events That Shaped a Movement* (Arlington, Virginia: Citizen's Clearinghouse for Hazardous Wastes, Inc., 1984).

35. Ibid.

36. Ibid.; Michael Weisskopf, "EPA to Complete Love Canal Cleanup but Habitability Remains Uncertain," *Washington Post*, October 27, 1987; Eric Schmitt, "Axelrod Says 220 Love Canal Families Can Return," *New York Times*, September 28, 1988.

37. Federal Emergency Management Agency (FEMA), "Superfund Relocation Assistance," Washington, D.C., 1985; Charles Robinson, FEMA, private communication, February 22, 1988.

38. FEMA, "Summary of Superfund Activity," unpublished memorandum, 1988; "8 From Contaminated Town Lose Illness Suit," *New York Times*, June 9, 1988; Michael Weisskopf, "Buyouts Replacing Cleanups as Remedy for Polluted Communities," *Washington Post*, September 3, 1987.

39. Weisskopf, "Buyouts Replacing Cleanups."

40. David Maraniss and Michael Weisskopf, "Jobs and Illness in Petrochemical Corridor," *Washington Post*, December 22, 1987.

41. Ibid.

42. Mike Leary, "Poisoned Environment Worries Eastern Europe," *Philadelphia Inquirer*, October 4, 1987. See also Hilary F. French, "Industrial Wasteland," *World Watch*, November/December 1988.

43. "Soviet Environmental Official Urges Punishment for Polluters," *Journal of Commerce*, July 6, 1988.

44. UNEP, "Hazardous Chemicals."

45. Jon Nordheimer, "Dioxin's Effects in Italy Less Severe Than Had Been Feared," *New York Times*, January 31, 1983; Michael H. Brown, *The Toxic Cloud* (New York: Harper & Row, 1987).

46. Thomas W. Netter, "Dioxin of '76 Italian Accident Reported Destroyed," *New York Times*, July 14, 1986; Nordheimer, "Dioxin's Effects."

47. UNEP, "Hazardous Chemicals."

48. Jane H. Ives, ed., *The Export of Hazard* (Boston: Routledge & Kegan Paul, 1985); Postel, *Defusing the Toxics Threat*.

49. UNEP, "Hazardous Chemicals"; Michael Isikoff, "Twice Poisoned," *The Washington Monthly*, December 1987.

50. Christopher Flavin, *Reassessing Nuclear Power: The Fallout From Chernobyl*, Worldwatch Paper 75 (Washington, D.C.: Worldwatch Institute, March 1987); George M. Woodwell, "Chernobyl: A Technology That Failed," *Issues in Science and Technology*, Fall 1986.

51. Flavin, *Reassessing Nuclear Power*.

52. Ibid.

53. Wendy Grieder, U.S. Environmental Protection Agency (EPA), quoted in Nathaniel Sheppard Jr., "U.S. Companies Looking Abroad for Waste Disposal," *Journal of Commerce*, July 20, 1988.

54. Blaine Harden, "Outcry Grows in Africa Over West's Waste-dumping," *Washington Post*, June 22, 1988; Steven Greenhouse, "Toxic Waste Boomerang: Ciao Italy!" *New York Times*, September 1, 1988.

55. "Waste Imports Alarm Lebanese," *Journal of Commerce*, June 27, 1988; Harden, "Outcry Grows."

56. Hilary F. French, "Combating Toxic Terrorism," *World Watch*, September/October 1988.

57. Tom Goemans and Pier Vellinga, "Low Countries and High Seas," presented to the First North American Conference on Preparing for Climate Change: A Cooperative Approach, Washington, D.C., October 27–29, 1987; J.E. Prins, *Impact of Sea Level Rise on Society* (Delft, Netherlands: Delft Hydraulics Laboratory, 1986).

58. Robert C. Cowen, "Man-made Gases Increase the Chance of Major Weather Change," *Christian Science Monitor*, June 30, 1988; warming projections from U.S. National Academy of Sciences, *Changing Climate*, Report of the Carbon Dioxide Assessment Committee (Washington, D.C.: National Academy Press, 1983).

59. James G. Titus, EPA, "Causes and Effects of Sea Level Rise," presented to the First North American Conference on Preparing for Climate Change: A Cooperative Approach, Washington, D.C., October 27–29, 1987; Erik Eckholm, "Significant Rise in Sea Level Now Seems Certain," *New York Times*, February 18, 1986; Ann Henderson-Sellers and Kendall McGuffie, "The Threat from Melting Ice Caps," *New Scientist*, June 12, 1986.

60. Titus, "Causes and Effects."

61. Maumoon Abdul Gayoom, speech before the 42nd Session of the U.N. General Assembly, New York, October 19, 1987.

62. John D. Milliman et al., "Environmental and Economic Impact of Rising Sea Level and Subsiding Deltas: The Nile and Bengal Examples," unpublished paper, Woods Hole Oceanographic Institution, Woods Hole, Massachusetts, 1988.

63. Milliman et al., "Environmental and Economic Impact"; Daniel Jean Stanley, "Subsidence in the Northeastern Nile Delta: Rapid Rates, Possible Causes, and Consequences," *Science*, April 22, 1988.

64. Milliman et al., "Environmental and Economic Impact." For further information see Clyde Haberman, "A Steamy, Crowded Bangkok is Sinking Slowly Into the Sea," *New York Times*, May 1, 1983.

65. Milliman et al., "Environmental and Economic Impact"; Prins, *Impact of Sea Level Rise*.

66. Milliman et al., "Environmental and Economic Impact."

67. Ibid.

68. Ibid.

69. Ibid.

70. Ibid.

71. Ibid.

72. Ibid.

73. Ibid.; PRB, "1988 World Population."

74. Milliman et al., "Environmental and Economic Impact."

75. Ibid.

76. Ibid.

77. Ibid.

78. Ibid.; PRB, "1988 World Population."

79. Milliman et al., "Environmental and Economic Impact."

80. Prins, *Impact of Sea Level Rise*; Gayoom, speech before the General Assembly.

81. Goemans and Vellinga, "Low Countries and High Seas"; U.S. coastline figure from *The 1988 Information Please Almanac* (New York: Houghton Mifflin Co., 1987).

JODI L. JACOBSON is a senior researcher at Worldwatch Institute. She is author of Worldwatch Paper #80, *Planning the Global Family*. She is coauthor of several other Worldwatch papers on demographic and social issues, and of the 1987, 1988, and 1989 editions of *State of the World*. Ms. Jacobson is a graduate of the University of Wisconsin in Madison, where she studied economics and environmental science.

THE WORLDWATCH PAPER SERIES

*Worldwatch Papers 2, 4, 5, 6, 8, 9, 11, 12, 13, 14, 15, 17, 19, 20, 22, 23, 24, 26, 27, 32, and 37 are out of print.

_____ **Total Copies**

Bulk Copies (any combination of titles)   **Single Copy** $4.00
    2–5: $3.00 each   6–20: $2.00 each   21 or more: $1.00 each

Calendar Year Subscription (1989 subscription begins with Paper 88)   U.S. $25.00 ___

Make check payable to Worldwatch Institute
1776 Massachusetts Avenue, N.W., Washington, D.C. 20036 USA

Enclosed is my check for U.S. $ _____

_____
name
_____
address
_____
city   **state**   **zip/country**